便便和树

From Poo to Tree

Gunter Pauli

冈特·鲍利 著
凯瑟琳娜·巴赫 绘
郭光普 译

www.xuelinpress.com

丛书编委会

主　任：贾　峰

副主任：何家振　闫世东　郑立明

委　员：牛玲娟　李原原　李曙东　李鹏辉　吴建民
　　　　彭　勇　冯　缨　靳增江

特别感谢以下热心人士对译稿润色工作的支持：

王必斗　王明远　王云斋　徐小怗　梅益凤　田荣义
乔　旭　张跃跃　王　征　厉　云　戴　虹　王　逊
李　璐　张兆旭　叶大伟　于　辉　李　雪　刘彦鑫
刘晋邑　乌　佳　潘　旭　白永喆　朱　廷　刘庭秀
朱　溪　魏辅文　唐亚飞　张海鹏　刘　在　张敬尧
邱俊松　程　超　孙鑫晶　朱　青　赵　锋　胡　玮
丁　蓓　张朝鑫　史　苗　陈来秀　冯　朴　何　明
郭昌奉　王　强　杨永玉　余　刚　姚志彬　兰　兵
廖　莹　张先斌

目录

便便和树	4
你知道吗?	22
想一想	26
自己动手!	27
学科知识	28
情感智慧	29
艺术	29
思维拓展	30
动手能力	30
故事灵感来自	31

Contents

From Poo to Tree	4
Did you know?	22
Think about it	26
Do it yourself!	27
Academic Knowledge	28
Emotional Intelligence	29
The Arts	29
Systems: Making the Connections	30
Capacity to Implement	30
This fable is inspired by	31

一棵苹果树正在努力结出果实,但这么多年来,土壤一直很贫瘠,并且缺少微生物。一只停在树枝上的冠蓝鸦感觉到了苹果树的压力。

"让我给土壤加点鸟粪吧。"冠蓝鸦建议道,"微生物会喜欢的。"

"多谢你的好意。但这不光是养料的事,我们还需要有机质。"苹果树回答。

An apple tree is making a great effort to grow fruit, but after all these years the soil is poor and the bacteria are scant. A blue jay sitting on a branch can tell the tree is stressed.

"Let me add some bird poo to the soil," the blue jay offers. "The bacteria will enjoy this."

"Thank you so much for your effort. But it is not only about food, we also need carbon," the tree responds.

一棵苹果树的压力……

An apple tree is stressed ...

有机质能滋养土地……

Carbon feeds the soil …

"有机质能做些什么呢？"

"有机质能滋养土地。"苹果树说，"土壤中的有机质越多，能生产的食物就越多。"

"这么说你真的不需要我的粪便吗？"小鸟问道。

"What is carbon good for?"
"Carbon feeds the soil," says the apple tree. "The more carbon in the soil the more food is produced."
"So you really do not need my poo?" asks the bird.

"当然不是，我同样需要你的粪便。而且我还需要能留住水分的土壤。"

"这也是有机质的功劳吗？"

"是的，有机质是很关键的。从土壤中拿走了多少有机质，就必须要返回更多才行！"苹果树解释道。

"Of course, I need your poo as well. And I need a soil that can retain water."

"Is that also thanks to the carbon?"

"Oh yes, carbon is key. A lot more has to return to the soil as so much has been taken out!" the tree explains.

我同样需要你的粪便。

I need your poo as well.

为了金属而进行地下采矿……

Mining the underground for metals ...

"听上去好像我们正在开采土壤中的矿物质来制造维生素,就像我们为了金属甚至黄金而进行地下采矿一样。"冠蓝鸦指出。

"太对了,我们'开采'土壤以获得有机质和矿物质来制造食物。如果我和我以后的树想一代代地生产富有营养的苹果,我们要把所有剩余的东西都还给土壤。我们想要的比得到的多得多。"

"It sounds like we are mining the soil for minerals to make vitamins, just like we are mining the underground for metals, and even gold," the blue jay observes.

"So right, we are mining soil for carbon and minerals to produce food. We need to have all the leftovers put back if I, and the trees that come after me, are to produce apples with a lot of goodies for generations to come. We need a lot more than we are getting today."

"那么谁会把这些营养还给你呢？"冠蓝鸦问道。

"首先，就是拿走那些养分的人。"苹果树马上回答。

"人类？人类有什么可以添加到土壤中的？"

"And who could get all these goodies back to you?" asks the blue jay.

"In the first place, the ones who took it out," the apple tree responds quickly.

"People? What do people have to add to the soil?"

人类有什么可以添加到土壤中的？

What do people have to add to the soil?

用塑料制品包裹婴儿的屁股……

Wrapping baby bums in plastics …

"告诉你吧,有很多的。尤其是他们的孩子。他们有很多粪便,绝对不应该丢到垃圾场。"

"我看到了他们现在的生活习惯!他们用填充了树木纤维的塑料制品包裹婴儿的屁股,婴儿排便后就直接丢掉了。"冠蓝鸦说。

"为什么不使用生物塑料和竹纤维制品呢?我想知道为什么这些混合物不还给我。"苹果树问。

"They have a lot, I tell you. Especially their babies. They have a wealth of poo that should never end up in a landfill."

"I have seen this modern day habit of theirs! Wrapping baby bums in plastics filled with tree fluff – and when it is soiled then it is all simply thrown away," Blue Jay says.

"Why are there no bioplastic used, and no bamboo fluff? I want to know why this mix does not come back to me," Apple Tree asks.

"你只想要婴儿刚拉出的那些东西吗？"

"不，我更希望再加上些厨余垃圾。而且绝对不要任何化学添加，但一定要加上些竹炭。"

"竹炭？"冠蓝鸦疑惑地说。

"You want that stuff just like that, fresh from the baby's bottom?"

"No, I would like it with some kitchen waste. And certainly without any chemicals, but definitely with some bamboo charcoal."

"Bamboo charcoal?" questions Blue Jay.

竹炭？

Bamboo charcoal?

……感谢土壤微生物……

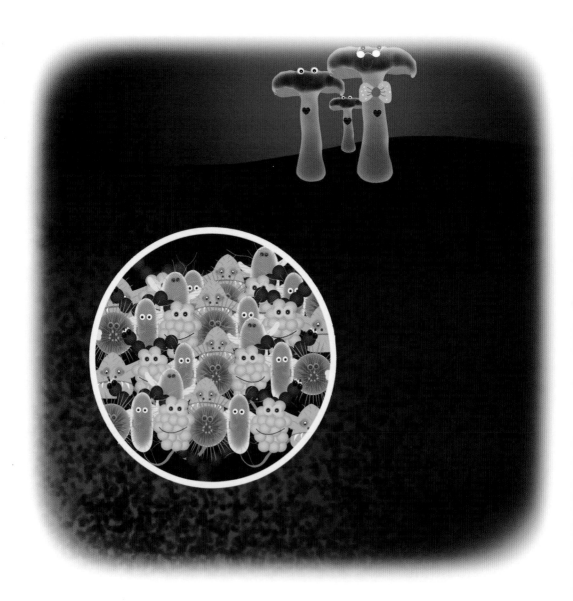

...thanks to bacteria and fungi...

"当然，竹子的剩余产品可以做成竹炭，这对皮肤非常好。婴儿需要它，而且我也爱它。"

"你给自己列了一张食谱。如果我没记错的话，食谱上有：生物塑料、厨余垃圾、竹纤维，还有竹炭。"

"别忘了还有婴儿粪便；感谢土壤微生物用它制造出我需要的黑土地。"苹果树补充道。

"Of course, bamboo left-overs are turned into charcoal and that is so good for your skin. Babies need it, and I love it."

"You got yourself quite a recipe. If I remember correctly it is: bioplastic, kitchen waste, bamboo fluff, and bamboo charcoal."

"And do not forget the baby poo; that will create the black soil I need, thanks to bacteria and fungi," Apple Tree adds.

"人们为啥非要自找麻烦地把这些给你呢?"

"是这样,感谢婴儿们的捐赠,黑土地上生长的果树可以生产出足够全城人吃的水果。"

"很难想象还有比这更划算的交易了。"冠蓝鸦惊讶地嘀咕着。

……这仅仅是开始!……

"Why should people even bother giving this to you?"

"Well, thanks to what the babies donate, fruit trees that grow in black soil will be able to produce so much fruit that everyone in the whole city will have enough to eat."

"It is hard to imagine a better deal," the astounded Blue Jay whispers.

... AND IT HAS ONLY JUST BEGUN!...

……这仅仅是开始！……

... AND IT HAS ONLY JUST BEGUN! ...

Did You Know?
你知道吗？

Soil organic carbon is the amount of carbon stored in the soil and is the basis of soil fertility. It is also a buffer against harmful substances. Good soils have +10% organic carbon, while with exploited soils it is likely to be less than 1%.

土壤有机碳是土壤中的含碳量，是土壤肥力的基础。它也有对抗有害物质的缓冲作用。优质的土壤中含有10%的有机碳，而被开发过的土壤中则低于1%。

All carbon in the soil combined store twice the amount of carbon in the atmosphere and vegetation. The carbon cycle starts with plants using carbon dioxide to produce organic matter. When plants and animals decompose minerals are released into the soil.

土壤中的含碳量是大气和植被中含碳量的两倍。碳循环开始于植物用二氧化碳制造有机物。当动植物被分解时，矿物质就被释放到土壤中。

Soil carbon levels have dropped by up to half of pre-agricultural levels as a result of fallowing, cultivation, burning and overgrazing. This means that there is a great deal of soil that have the potential for a large increase in soil organic carbon.

由于抛荒、种植、燃烧和过度放牧，土壤中的碳水平比农业开始前降了一半。这意味着有大量的土壤有潜力大幅提升有机碳含量。

Carbon storage is a free yet indispensable ecosystem service. Soil carbon improves the retention of water, reduces the need for irrigation and cuts the risk of flooding.

碳储备是一种免费却不可或缺的生态系统服务。土壤中的碳可以增加保水能力，减少对灌溉的需求，并降低洪水风险。

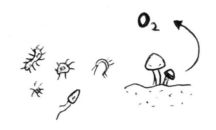

Soil organic matter is composed of bacteria and fungi and decaying tissue and fecal matter material. It is the result of photosynthesis, respiration and decomposition.

土壤有机质由细菌和真菌、腐殖质和粪便组成。它是光合作用、呼吸作用和分解作用的结果。

Disposable diapers used in the 21st century will only finish degrading by the year 2500. It takes four centuries for this mix of plastics and human waste to turn to soil.

21世纪使用的一次性尿布要到2500年才能降解完。这种混合了塑料和人体废物的垃圾需要四个世纪才能转化为土壤。

Apples are less nutritious today than 50 years ago. Modern intensive agriculture has stripped nutrients from the soil, and selected seeds of fruits, vegetables and grains offer only volume of output and not quality content.

现在的苹果不如50年前的有营养。现代集约型农业从土壤中掠夺了营养物质，所选育出的水果、蔬菜和谷物品种只是产量大，营养质量并没有提升。

There is one traditionally farmed apples species that has a 100 times more nutrients than the Golden Delicious available in supermarkets. There are 2 500 varieties of apples grown in the USA but only the crabapple is a native.

有一种传统种植的苹果品种的营养成分比超市中的"金美味"苹果高出100倍。美国种植着2500种苹果，但只有沙果是本地品种。

Think About It 想一想

Would you prefer to eat a lot of food with little nutritional content, or eat a small volume of food with lots of nutrients?

你更喜欢吃大量低营养的食物,还是少量高营养的食物呢?

Can we farm in such a way so that carbon gets put back into the soil?

我们能用某种方式耕作以便让更多的碳回到土壤里吗?

Should we suggest that no disposable diapers are used for babies, or should we do something useful with the soiled diapers?

我们是应该建议不要给婴儿使用一次性尿布,还是应该用脏尿布做一些有用的事情?

Does it make sense to ship bamboo and its charcoal around the world?

在世界各地运输竹子及竹炭是否有意义?

Do It Yourself!

自己动手！

Do you eat apples? Do you know which varieties of apples are from your part of the world? Apples grow almost anywhere. Learn how to choose the best apple tree to plant in your garden, or on a rooftop or balcony. If you plant apple trees in the city, go for a hedge of a dwarf apple trees. Some apples ripen early in the season and are ideal for making apple sauce, while apples that ripen later are far better for storage, so you can eat them over a long period of time. Of course, check which apple variety is rich in the type of vitamins you need so that you can tailor it to your nutritional needs to maintain good health.

你吃苹果吗？你知道世界上什么苹果品种产自你所生活的地方吗？几乎所有地方都能生长苹果。学习如何选择最好的苹果树种在你的花园、屋顶或者阳台上。如果你要在城市种苹果树，可以用矮苹果树做个树篱。早熟型的苹果适合制作苹果酱，而晚熟型的苹果更适合储存，可以让你吃很长时间。查下哪种苹果能提供你所需要的丰富的维生素种类，将它放入保持健康的营养需求之中。

TEACHER AND PARENT GUIDE

学科知识
Academic Knowledge

生物学	苹果属于蔷薇科；苹果树是一种落叶乔木；苹果的膳食纤维在健康饮食中的作用；动植物组织降解的重要性；堆肥和黑土之间的区别。
化 学	苹果中不含脂肪、钠或胆固醇；苹果籽中含有氰化物和蓖麻酸；当苹果破损或被切开后，果肉会氧化变成棕色；质量差的刀具所含的铁盐会促使切开的苹果变成棕色；碘可以用于检测苹果和马铃薯中的淀粉含量，如果苹果变成紫色，意味着淀粉含量高。
物 理	防止苹果"生锈"可以通过加热、降低pH（使用柠檬汁）和减少氧气（真空包装）实现；一次性尿布是以液体吸收为核心的现代发明。
工程学	新鲜采摘的苹果被粉碎和制浆、压榨提取果汁、添加酵母发酵将糖转化成酒精，然后在桶中贮藏数月至醇熟。
经济学	苹果种植是劳动密集型产业，大多数苹果是手工采摘的；经济学的选择是由产量（更多的苹果）驱动的，而不是基于其营养含量。
伦理学	我们如何评价从土壤中获取营养物质而不给予补充？这相当于开采土壤；在将营养物质还给土壤之前，我们怎么能允许使用需要400年才能降解的产品？
历 史	在亚瑟王的故事中，神圣的岛屿"Avalon"被翻译为"苹果岛"；城市下水道还没发明前，人们将夜土（即人类粪便）收集起来运出城镇，给农田施肥或者直接倒入附近的河流；安第斯山脉和亚马逊的文化是如何生产黑土来创造和维持其土壤肥力的。
地 理	哈萨克斯坦的城市阿拉木图的名字来源于哈萨克语"苹果"，因此经常被翻译成"满地苹果"。
数 学	倍增效应：1000个婴儿一年产生的废物足以给1000棵果树施肥，如果10年后每棵树每年生产400公斤苹果，那么25年后城市里就将充满新鲜水果。
生活方式	现在的社会时兴丢弃，所有的产品都是一次性的，造成了大量的浪费（和大量的成本），也打破了碳循环的平衡。
社会学	如果我们不补充土壤肥力，还把最好的天然原料（婴儿粪便）直接丢弃掉，社会如何实现食品供应的持续性；艾萨克·牛顿受到苹果从树上掉落的启发而发现了引力定律；"一天一苹果，医生远离我"，这个俗语督促我们多吃新鲜水果。
心理学	婴儿刚学会走路时，他们会对又湿又冷又不舒服的脏尿布很敏感，这有助于他们进行如厕训练；当超吸收物让婴儿感到温暖、干燥、没有任何不适时，他们会需要更多的时间来学习基础的卫生习惯。
系统论	我们需要补充土壤，因为没有任何遗传工程能最终替换有机碳循环。

教师与家长指南

情感智慧
Emotional Intelligence

冠蓝鸦

冠蓝鸦表现出对树的同情，并准备帮助他。冠蓝鸦对自己和树的关系有信心，他提出问题以便更好地了解新的信息。冠蓝鸦的思考过程严谨，他善于分析，还用采矿业作比较。他希望参与到事件的解决方案中，但是指出首先需要认识到人类的生活方式存在着严重问题，尤其是在一次性卫生用品的使用上。苹果树提出了一种新奇的解决方案，而冠蓝鸦想要确定自己对这个新产品设计方案的每个方面都能理解。在向他解释整个系统之前，他还是很难想象为什么人们会选择利用黑土地。

苹果树

苹果树很感谢冠蓝鸦，并向他详细解释了他需要什么，并表明这个解决方案中包含鸟粪。接着，苹果树将短期问题与长期解决方案联系起来形成一个基于重新启动碳循环的解决方案。苹果树纳闷为什么婴儿的排泄物没有混合入厨余垃圾和木炭然后再送回到土壤中。苹果树耐心地深入探讨了细节，解释了如何在微生物例如细菌和真菌的帮助下制造黑土壤。他说服了冠蓝鸦，这个解决方案除了单纯的物质交换外还有极大的好处，即如果这个措施能实施的话，那么每个人都能免费吃到水果。

艺术
The Arts

准备一个苹果、一堆黑土、水彩颜料、画笔、水和纸。首先通过触摸探索、感受和挤压来了解苹果，观察它的形状和颜色。接下来用手接触并感觉早晨在森林里收集的新鲜黑土。现在拿起画笔和纸，画出你对苹果和黑土的体验，讲个故事说说你在亲密接触它们时的感受。记住要有创意，画中的颜色不必与苹果和土壤的色彩一致！

TEACHER AND PARENT GUIDE

思维拓展
Systems: Making the Connections

　　健康又富含矿物质和微生物的表层土壤是为地球上所有生物提供水和食物的关键。表层土壤的更新和补给是经常被忽视的生态系统服务。最近的观察显示，水果、蔬菜、豆类和坚果营养浓度不及以前的作物，这主要是因为我们没有为表层土壤补充养分。人类像开采黄金一样正开采并耗尽土壤中的矿物质。如果没有生物的多样性和强度，植物就不能合成我们需要的丰富营养。然而，现代设计和生产的产品变成了废物生产线，这些产品含有的几个世纪才能降解的塑料以及化学处理的纤维素堆积在垃圾填埋场。一次性尿布，这种让婴儿感到舒适的"包裹屁股的绷带"会使孩子继续弄脏自己而无法产生学习使用厕所的迫切需求。将人类、动物和植物产生的废弃物返还给土壤的循环正在被破坏，而世界各地的表层土壤再也不能仅靠肥料补充了。基因工程的目的只是高产量，而更大的需求是提高营养含量。高山苹果的营养含量是超市苹果的百倍，秘鲁高地的马铃薯品种含有的抗癌物质比普通品种多28倍。当我们发现这些时，我们认识到两种趋势的结合导致产品营养缺乏。再加上无节制地工业化养殖和储备食物，同时可用的土壤被进一步开采——常常使碳含量低于1%。土壤被侵蚀，保水性下降，生产能力下降，我们只能通过极端的基因工程和大量使用化肥来暂缓这种下降。保持食物和水长久安全的关键是保持食物和营养的良性循环，比如启动黑土壤的休养周期，以丰富土壤含碳量。

动手能力
Capacity to Implement

　　让我们学习如何使用新鲜水果来制作十分美味和健康的零食。烤苹果片是油腻炸薯片的绝佳替代。你可以这样来制作自己的水果干：将水果切片在食物脱水器中放几小时，或在温热的烤箱中过夜。将苹果片浸入柠檬汁中来防止变色。将地姜加入菠萝会有特殊的风味。可以在苹果片上加入坚果、燕麦和葡萄干，再加些苹果糖浆和核桃油。与你的朋友分享这些健康小零食，鼓励别人加入热爱新鲜水果的行列！

教师与家长指南

故事灵感来自

松井松美
Ayumi Matsuzaka

松井松美出生于日本长崎，在东京的日本大学完成艺术学习后，前往德国魏玛的包豪斯大学学习，从此就居住在那里。松井松美是一位行动艺术家和社会企业家，将人体与自然循环联系起来。她热衷于将有机废物，比如头发、指甲和自己的身体废物转化为土壤成分。她受邀在博物馆和画廊向好奇的观众分享她的知识和技能，2015年她从艺术领域拓展到企业。她是"DYCLE – 尿布循环"的创始人，该公司用废弃尿布来制造土壤成分，是一家致力于种植数百万棵树的开源企业。

图书在版编目（CIP）数据

便便和树：汉英对照/（比）冈特·鲍利著；（哥伦）凯瑟琳娜·巴赫绘；郭光普译．— 上海：学林出版社，2017.10
（冈特生态童书．第四辑）
ISBN 978-7-5486-1234-6

Ⅰ．①便… Ⅱ．①冈… ②凯… ③郭… Ⅲ．①生态环境－环境保护－儿童读物－汉、英 Ⅳ．① X171.1-49

中国版本图书馆 CIP 数据核字（2017）第 133796 号

© 2017 Gunter Pauli
著作权合同登记号　图字 09-2017-532 号

冈特生态童书
便便和树

作　　　者——	冈特·鲍利
译　　　者——	郭光普
策　　　划——	匡志强　张　蓉
特约编辑——	李玉婷
责任编辑——	许苏宜
装帧设计——	魏　来
出　　版——	上海世纪出版股份有限公司 学林出版社
	地　址：上海钦州南路 81 号　电话/传真：021-64515005
	网　址：www.xuelinpress.com
发　　行——	上海世纪出版股份有限公司发行中心
	（上海福建中路 193 号　网址：www.ewen.co）
印　　刷——	上海丽佳制版印刷有限公司
开　　本——	710×1020　1/16
印　　张——	2
字　　数——	5 万
版　　次——	2017 年 10 月第 1 版
	2017 年 10 月第 1 次印刷
书　　号——	ISBN 978-7-5486-1234-6/G.460
定　　价——	10.00 元

（如发生印刷、装订质量问题，读者可向工厂调换）